The Qualitative Estimation of BCR-ABL Transcript

(An In-Lab Procedural Study on Leukemia Patients)

Hitesh Goyal

ISBN 978-93-5610-294-1
© Hitesh Goyal 2022
Published in India 2022 by Pencil

A brand of
One Point Six Technologies Pvt. Ltd.
123, Building J2, Shram Seva Premises,
Wadala Truck Terminal, Wadala (E)
Mumbai 400037, Maharashtra, INDIA
E connect@thepencilapp.com
W www.thepencilapp.com

Author biography

Hitesh Goyal is the author of this manuscript. Mr. Goyal has earned his graduation in Applied Biotechnology with a specialization in medical biotechnology. Mr. Goyal has done his internship at the Department of Molecular Genetic Labs under Sir Ganga Ram Hospital, New Delhi. Also, he has done a small project for the presentation of his dissertation upon graduation at Raj Rishi Govt. College under Rajasthan University (earlier) and now under Matsya Brihtari University, Alwar, Rajathan.

Goyal has been a meritorious student in his graduation studies. Goyal's other work in graduation such as Quality estimation of water at Raj Rishi College Campus, Alwar was one of the good studies under the supervision of Applied Green Chemistry Lab, Raj Rishi College, Alwar, Rajasthan.

CONTENTS

Preface

The BCR-ABL gene shows up in patients with certain types of leukemia, a cancer of the bone marrow and white blood cells. BCR-ABL is found in almost all patients with a type of leukemia called chronic myeloid leukemia (CML). Another name for CML is chronic myelogenous leukemia. Both names refer to the same disease.

The BCR-ABL gene is also found in some patients with a form of acute lymphoblastic leukemia (ALL) and rarely in patients with acute myelogenous leukemia (AML).

Certain cancer medicines are especially effective in treating leukemia patients with the BCR-ABL gene mutation. These medicines also have fewer side effects than other cancer treatments. The same medicines are not effective in treating different types of leukemia or other cancers. Other names: BCR-ABL1, BCR-ABL1 fusion, Philadelphia chromosome.

This book addressed a very short study on the basis of a graduation dissertation conducted on the drug imatinib generally used for the treatment of leukemia cancer. The research studies have administered the impact of the drug imatinib and others on the problem of leukemia, especially in the patients with genetic cancer affected by leukemia by the Philadelphia chromosome issue.

Acknowledgements

I would like to acknowledge all the references I have quoted in this book. Also, I am thankful to my professors who gave me an opportunity to publish this manuscript in book format.

I am also thankful to the publisher for giving me chance to publish my work in a book format.

Introduction

Leukemia

Leukemia is a group of neoplastic disorders that arises in the hematopoietic cells of the bone marrow and leads to an uncontrolled proliferation and accumulation of immature blood cells. This excessive production of blood cells leads to overcrowding of the bone marrow and spreading into the peripheral blood and to other organs. The lack of functional blood cells can lead to symptoms like anemia, infections, and bleeding. If left untreated, leukemia is fatal, often due to complications resulting from the leukemic infiltration of the bone marrow and replacement of normal hematopoietic precursor- cells (Herfindal et al, 2000).

The first published description of a case of leukemia in medical literature dates to 1827, when French physician Alfred-Armand-Louis-Marie Velpeau described a 63-year-old florist who developed an illness characterized by fever, weakness, and urinary stones, and substantial enlargement of the liver and spleen. Velpeau noted that the blood of this patient had a consistency "like gruel", and speculated that the appearance of the blood was due to white corpuscles. In 1845, a series of patients who died with enlarged spleens and changes in the "colors and consistencies of their blood" was reported by the Edinburgh-based pathologist J.H. Bennett; he used the term "leucocythemia" to describe this pathological condition (Bennett et al, 1845). The term "leukemia" was coined by Rudolf Virchow, the renowned German pathologist, in 1856. As a pioneer in the use of the light microscope in pathology, Virchow was the first to describe the abnormal excess of white blood cells in patients with the clinical syndrome described by Velpeau and Bennett. As Virchow was uncertain of the cause of the white blood cell excess, he used the purely descriptive term "leukemia" (Greek: "white blood") to refer to the condition (Virchow, 1856).

Leukemia according to the predominant cell type involved (Henderson et al, 2002). Traditionally, leukemia was classified as chronic or acute by how fast the disease progressed to a fatal clinical outcome. This has later been found to correlate well with the degree of maturation of the predominant malignant cell. The disease is further classified into lymphoid or myeloid.

This book is based on the findings of the study conducted by a researcher on patients with leukemia in India, especially in the Delhi Region. The research has covered the patients as participants from across the country and even foreign too.

Review of Literature

Chronic myelogenous leukemia

Chronic myelogenous leukemia (CML) is a clonal myeloproliferative disorder of the pluripotent stem cell. The disease is usually characterized by the insidious onset of symptoms, progressive splenomegaly, marrow hypercellularity, anemia, leukocytosis, and cytogenetically by the presence of Philadelphia (Ph) chromosome t (9; 22) (q34; q11) in 90 to 95% of patients.

The disease follows a biphasic or triphasic course. There is an initial chronic phase which after an average period of 5 to 5.5 years may progress to an intermediate phase called the accelerated phase followed by blastic transformation or blast crisis. About 90-95% of patients present in the chronic phase at diagnosis, remaining 5-10% may have features of advanced disease (Sawyers et al, 1999).

Chronic myeloid leukemia was probably the first form of leukemia to be recognized as a distinct entity. In 1845, two patients were described as having massive splenomegaly associated with leukocytosis, which seemed to be a novel entity not explained by the other causes of splenomegaly, such as tuberculosis, that were already widely accepted in the 1840s (Geary et al, 2000).

The first important clue to its pathogenesis came only very much later when in 1960 newly developed techniques for studying human cells in mitosis allowed Nowell and Hungerford to detect a consistent chromosomal abnormality, termed as Philadelphia (Ph) chromosome and identified as 22q-, in patients with this disease (Nowell and Hungerford, 1960).

In 1973, Rowley observed that the Ph chromosome resulted from a reciprocal translocation that involved chromosomes 9and 22; the abnormality was then designated as t (9; 22) (q34; q11) (Rowley, 1973).

In the 1980s, the Ph chromosome was shown to carry a unique fusion gene, termed BCR-ABL, the generation of which is now believed to be the principal cause of the chronic phase of CML (de Klein et al, 1982; Collins et al, 1984; Gale et al, 1984; Shtivelman et al, 1985).

Incidence Status

The incidence of CML is 1 per 100,000 population in the West (Parkin et al, 2005). The true incidence of CML in India is not available. According to six population-based Cancer Registries (covering <0.3% of total population), the incidence of CML in India varies from 0.8 to 2.2 per 100,000 population for males and from

0.6 to 1.6 per 100,000 population in females (National Cancer Registry).

Hospital-based studies have reported a higher frequency of CML ranging from 40% to 82% of all cases of leukemia among adults (Bhutani et al, 2002).

Clinical and Hematological Features

The median age of onset is 38 to 40 years in India compared to 50 years in the West (Bhutani et al, 2002). There is a slight male preponderance. With routine screening tests, 5-15 % of patients are diagnosed in the asymptomatic stage. The presenting symptoms are usually malaise/ fatigue, abdominal fullness, fever, weight loss, abdominal pain, and occasionally easy bruising or bleeding. Splenomegaly is usually present in 90% of patients and in one-third of them it is more than 10 cm. nearly 25% of patients have significant hepatomegaly (>2cm), lymphadenopathy is uncommon (<10%) in the chronic phase- CML and is confined to 1-2 regions with small nodes. The initial hematological parameters are - Hemoglobin >10G/dl, total leukocyte count (WBC) 100-300 x 109/L, Platelets- 200-400x109 /L. On differential leukocyte count (DLC) myeloid series of cells are seen in all stages of maturation. Basophils are elevated but only 10% to 15% of patients have [3] 7% basophils in peripheral blood. Frequently, eosinophils are mildly elevated. The bone marrow (BM) is hypercellular and devoid of fat.

There is myeloid hyperplasia with a myeloid to erythroid ratio usually 10-30:1. On differential leukocyte count - all stages of maturation of myeloid series are usually seen with myeloblasts and promyelocytes <10% and with the predominance of myelocytes. Evidence of focal fibrosis may be seen on reticulin stain in 25-28% of patients in the chronic phase but increases with disease progression. The biochemical abnormalities include- low LAP (leucocyte alkaline phosphatase) score, marked elevation of serum B12, and the B12 binding protein transcobalamin-I. Hyperuricemia related to increased cell turnover may occur prior to therapy and may be exacerbated by treatment. LAP score may increase with infection, clinical remission, or onset of blast crisis (Kumar et al, 2006).

Accelerated Phase / Blast Crisis

With conventional treatment, the chronic phase progresses into an accelerated phase (AP) that lasts for 1-1.5 years and is followed by a blastic phase (acute phase). The transition to blast crisis (BC) may be abrupt in 20-25% of patients without an intermediate AP. BC is a terminal event in 70% of patients. (Giles et al, 2004).

World Health Organization (WHO) criteria defining accelerated phase and blast crisis are mentioned in Table 2.1. Occasionally, extramedullary blastic infiltrates in lymph nodes, bone or skin may precede BC in BM. Phenotypically blasts are mainly myeloblastic (60%), lymphoblastic (20%), and undifferentiated in 10-15% of patients. Rarely, there may be erythroblasts, megakaryocytic or mixed transformation (Giles et al, 2004).

Focal myelofibrosis may be seen in up to 30% of patients on presentation. Increasing myelofibrosis on serial BM biopsies may be associated with AP/BC (Giles et al, 2004).

*WHO Criteria used for CML staging (Baccarni M et al , Blood.2006;108:1809-1820)*1

Disease stage	Definition
Chronic phase (CP)	• <10% blasts in peripheral blood (PB) or bone marrow (BM) • <20%, basophils in blood • Platelets count >100 X109/L
Accelerated phase (AP)	• Blast cells in blood or bone marrow 10%-19% • Basophils in blood 20% or more • Persistent thrombocytopenia (platelet count less than 100 X

	109/L) unrelated to therapy • Thrombocytosis (platelet count greater than 1000 X 109/L) unresponsive to therapy • Increasing spleen size and increasing WBC count unresponsive to therapy • Cytogenetic evidence of clonal evolution (the appearance of additional genetic abnormalities that were not present at the time of diagnosis)
Blast crisis (BC)	• $\geq$ 20% blast cells in blood or bone marrow • Extramedullary blast proliferation, or large foci or clusters of blasts in the bone marrow biopsy.

The Philadelphia (Ph) Chromosome

CML was the first human cancer to be associated with a chromosomal abnormality, and its discovery was a breakthrough in cancer biology (Sawyers et al, 1999). The Philadelphia chromosome was first discovered and described in 1960 by Peter Nowell from the University of

Pennsylvania School of Medicine and David Hungerford from the Fox Chase Cancer Center's Institute for Cancer Research and was therefore named after the city in which both facilities are located (Nowell and Hungerford, 1960). Using quinacrine fluorescence and Giemsa banding, Rowley and colleagues showed that the Ph chromosome resulted from a reciprocal translocation between the long arms of chromosomes 9 and 22; t(9;22)(q34;q11) (Rowley, 1973). This is the hallmark of CML and plays a key role in CML pathogenesis (Figure 2.1). It is present in more than 95 percent of CML patients, also present in 5 percent of children and in 15 to 30 percent of adults with Acute Lymphoblastic Leukemia (ALL), and in 2 percent of patients with newly diagnosed acute myeloblastic leukemia ((Shepherd et al, 1995; Kurzrock et al, 1988; Specchia et al, 1995). This reciprocal translocation results in the transfer of the 3' segment of the ABL gene from chromosome 9q34 to the 5' part of the BCR gene on chromosome 22q11, creating a hybrid BCR–ABL gene that is transcribed into a chimeric BCR–ABL messenger RNA (mRNA) (de Klein et al, 1982; Collins et al, 1984; Gale et al, 1984; Shtivelman et al, 1985). BCR-ABL fusion gene generated in this reciprocal translocation codes for BCR-ABL chimeric protein which is a constitutively active tyrosine kinase enzyme (Konopka et al, 1984; Davis et al, 1985; Ben-Neriah et al, 1986; Mes-Masson et al, 1986; Lugo et al, 1990; Daley et al, 1990).

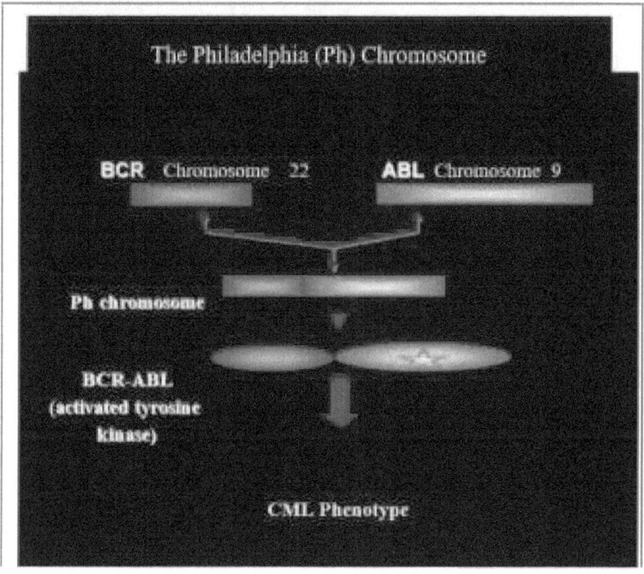

Philadelphia chromosome is formed as reciprocal translocation between chromosome 9 and 22 which results in the fusion of ABL gene on chromosome 9 to BCR sequences on chromosome 22.BCR-ABL fusion gene is central to CML pathogenesis and lead to CML phenotype.

The ABL gene contains 11 exons among which the first exon has two variants: 1a and 1b (Kurzrock et al, 1988). The ABL gene encodes a ubiquitously expressed, non-receptor tyrosine kinase with a molecular mass of 145 KD (p145). The isoforms of ABL, i.e. 1a and 1b, derive from alternative splicing of the first exon. The breakpoint in the ABL gene may occur within a region longer than 300 kb, but usually before exon 2. The ABL exons 2 to 11 (also called a2 to a11) are juxtaposed to the 5' part of BCR.

The major breakpoint cluster region (M-bcr) of the BCR gene on chromosome 22 is located between exon 12 and 16 (referred to as b1 to b5) and extends over 5.8 kb (Heisterkamp et al, 1991).Two fusion transcripts, e13a2 and e14a2 (b2a2 and b3a2, respectively), are created and both translate into a chimeric protein of 210 KD (p210) (Kurzrock et al, 1988) (Figure1.2). In 95% of BCR-ABL positive CML patients, the leukemic cells have either b2a2 or b3a2 transcripts, but in 5 percent of cases, alternative splicing events cause the expression of both fusion products (Melo, 1996).

The breakpoint in the minor breakpoint cluster region (m-bcr) results in a fusion transcript named e1a2, which codes a 190 KD protein (p190) (Chan et al, 1987; Clark et al, 1987) (Figure 2.2). e1a2 BCR-ABL transcript is rare in CML and is mainly seen in adults and children with Ph-positive ALL (Kurzrock et al, 1987). Breakpoints in the micro breakpoint cluster region (μ-bcr) create another fusion transcript (e19a2) which is translated into a 230 KD protein, (p230). e19a2 BCR-ABL transcript has been associated with neutropenic CML (Pane et al, 1996) and some rare cases of CML (Wilson et al, 1997).

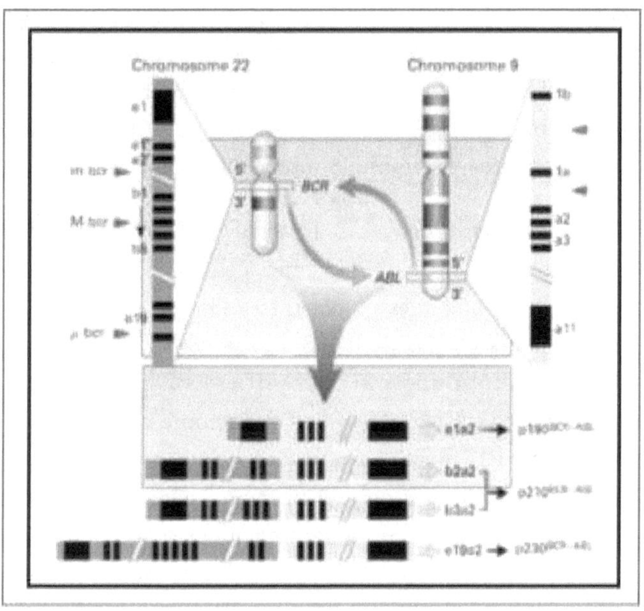

Philadelphia chromosome: BCR-ABL fusion transcript type

The Ph chromosome is a shortened chromosome 22 results from the translocation of 3' (toward the telomere) ABL segments on chromosome 9 to 5' BCR segments on chromosome 22. In most cases, breakpoints (arrowheads) in the ABL gene are located in the 5' end (toward the centromere) of exon a2. Various breakpoint locations have been identified along the BCR gene on chromosome 22. Depending on which breakpoints are involved, differently sized segments from BCR are fused with the 3' sequences of the ABL gene. This results in the fusion of messenger RNA molecules (e1a2, b2a2, b3a2, and e19a2) of different lengths that are translated into different chimeric protein products (p190, p210, and p230) with various molecular weights. m-bcr: minor breakpoint cluster region, M-bcr: major breakpoint cluster region, and μ-bcr: micro breakpoint cluster region. (Source: Faderl S, 1999)

Approximately, 3-10% of CML patients have cytogenetically normal leukemic cells (Ph negative). A proportion of these patients (30-80%) have re-arrangements of the BCR-ABL gene with the production of 8.5kb mRNA and p210 KD BCR-ABL protein similar to Ph-positive CML The clinical outcome of Ph negative but BCR-ABL positive, patients is similar to those with Ph-positive. Remaining Ph-negative patients with the absence of BCR-ABL re-arrangement have a distinct clinical course despite their early resemblance to classical CML. They eventually develop BM failure (anemia, thrombocytopenia) accompanied by markedly increased leukemia burden with increased WBC, organomegaly, and extramedullary disease. Blastic transformation generally does not occur. These are probably cases of myelodysplastic syndrome / chronic myelomonocytic leukemia. Since this sub-group is different from CML, World Health Organization (WHO) has defined this subgroup as "myeloproliferative syndrome, unclassifiable" (Hughes et al, 2006).

Cytogenetic and molecular genetic analysis

Philadelphia chromosome (Ph) resulting in the formation of the BCR-ABL fusion gene is the hallmark for CML. A conclusive diagnosis of CML relies on cytogenetic and/or molecular testing to identify this specific genetic abnormality. G-banding karyotyping is used for cytogenetic analysis (Moorhead et al, 1960) and usually 20-30 metaphase cells are analyzed. Cytogenetic analysis detects the Ph chromosome in approximately 95% of patients with CML at the time of diagnosis. Other remaining CML cases carry masked translocations that can be detected only by molecular techniques, such as fluorescence in situ hybridization (FISH) or reverse transcriptase-polymerase chain reaction (RT-PCR) for the BCR-ABL fusion gene (Cortes et al, 1995; Aurich et al, 1998).

Fluorescence in situ hybridization (FISH)

The FISH analysis is typically performed by co-hybridization of a BCR and an ABL fluorescent-labeled probe to denatured metaphase chromosomes or interphase nuclei. Traditional FISH (also known as S-FISH or dual-FISH) is a two-color technique in which a 5' BCR fluorescent probe and a second 3' ABL fluorescent probe are utilized with contrasting colors to detect the position of the respective genes (Tkachuk et al, 1990).

Reverse Transcriptase - PCR (BCR-ABL)

To monitor response in CML patients treated with newer therapeutic regimens that induce high response rates, a method with high sensitivity is needed. Reverse transcriptase PCR is one technique that can identify the presence or absence of BCR-ABL fusion transcripts. RNA isolated from peripheral blood samples is subjected to cDNA synthesis and then amplified using BCR-ABL fusion gene-specific primers and then visualized on an agarose gel under UV illumination. PCR techniques, such as RT-PCR, multiplex PCR, and nested-PCR, targeting the BCR-ABL fusion gene have been shown to detect the CML disease with high sensitivity and enabled more specific response assessment (van Dongen et al, 1999). But the serial measurement of leukemia-specific BCR-ABL transcript levels in the blood or bone marrow will be a more valuable approach for estimation of responses.

Qualitative PCR

The qualitative BCR-ABL test can determine the specific type (isoform) of the Philadelphia chromosome present which is important for appropriate diagnosis and treatment. Our qualitative BCR-ABL test detects the presence of the p190, p210, and p230 isoforms.

Treatment of CML

Imatinib mesylate

Imatinib mesylate (Signal Transduction Inhibitor (STI) - 571) or Gleevec is a molecularly targeted drug, approved by the Food and Drug Administration in May 2001 and has revolutionized the management of CML since then (Druker et al, 2001). In view of the high hematological and cytogenetic responses induced by Imatinib, it has now become the standard therapy for all phases of CML.

Imatinib Mesylate (STI-571)

As the BCR-ABL fusion protein plays a key role in CML pathogenesis, attempts to target the BCR-ABL tyrosine kinase evolved as new therapeutic strategies. The antecessor of imatinib was initially developed as a specific platelet-derived growth factor receptor (PDGFR) inhibitor at Ciba-Geigy (currently Novartis, Basel, Switzerland). Further, when optimized for v-ABL tyrosine kinase inhibition, imatinib mesylate was generated. (Buchdunger et al, 1995 (Traxler etal, 2001; Manley et al, 2002). Imatinib is a phenylaminopyrimidine derivative and selectively inhibits ABL tyrosine kinase, including BCR-ABL (Figure 2.3) (Manley et al, 2002; Buchdunger et al, 1996). Further studies revealed that imatinib can also target a few other tyrosine kinases including PDGFR (Buchdunger et al, 2000), c-KIT (Heinrich et al, 2000), and ARG (Okuda et al, 2001).

Preclinical studies showed that imatinib inhibits the proliferation of BCR-ABL (p210) positive cell lines and also inhibits the clonal growth of myeloid cells in CML patients (Druker et al, 1996; Deininger et al, 1997). Studies with mice models also revealed that imatinib had in vivo activity against BCR-ABL positive cells and continuous exposure to imatinib was necessary to eradicate the tumors (Druker et al, 1996; Le Coutre et al, 1999).

Imatinib was shown to have an acceptable toxicology profile in animal models, which was necessary before clinical testing.

Pre-experimental Planning and Methodology

The most obvious requirement before starting the designed work is to collect the material used in the process and to be familiar with the various methods and instruments required in the investigation Blood samples, related apparatus, and chemicals required for the experimental work were collected from the patient before starting the practical work.

- Blood samples: samples from five patient P1, P2, P3, P4, and P5 was collected from Cancer Hospital.

- Apparatus: micropipette, agarose gel electrophoresis, Centrifuge, PCR machines, Gel Doc (U.V. illuminator), PCR tubes, Eppendorf tubes(1.5 ml and 2 ml) and Pipette tips, stacker cooler or mini cooler, Centrifuge tubes (15ml and 50 ml), laboratory glassware.

- Chemicals: Nuclease free water, 5x C-DNA synthesis buffer, dNTPs, RNA primer, RT Enhancer, Verso Enzyme mix, Template (RNA),

Agarose, 0.5x buffer, ethidium bromide, bromophenol blue, molecular marker, Reverse transcriptase enzyme and DNA Taq Polymerase enzyme, RNAse inhibitor, Random Decamers, RNA isolation kit,

Methodology

Subject:

Three CML patients attending the outpatient department of genetic medicine, Hospital. Initial evaluation included – history,

- Sample collection: Peripheral blood and bone marrow samples of the patient were collected at the start of the Imatinib therapy as baseline samples.

Detection of BCR-ABL fusion transcript type

- Isolation of RNA from Qiagen kit:

Principle: QIAamp spin columns represent technology for total RNA preparation that combines the selective binding properties of a silica-based membrane with the speed and convenience of micro spin technology. A specialized high-salt buffering system allows RNA species longer than 200 bases to bind to the QIAamp membrane. During the QIAamp procedure for purification of RNA from blood, erythrocytes are selectively lysed and leukocytes are recovered by centrifugation. The leukocytes are then lysed using highly denaturing conditions that immediately inactivate RNases, allowing the isolation of intact RNA. After homogenization of the lysate by brief centrifugation through a QIA shredder spin column, ethanol is added to adjust binding conditions and the sample is applied to the QIA amp spin column. RNA is bound to the silica membrane during a brief centrifugation step. Contaminants are washed away and total RNA is eluted in 30 µl or more of RNase-free water for direct use in any downstream application. Since the procedure relies on intact leukocytes, frozen blood cannot be used. QIA amp RNA Blood Mini Handbook 04/2010 QIAamp RNA Blood Kits enrich for RNA species larger than 200 nucleotides since small RNAs such as 5.8S RNA, 5S RNA, and tRNA (approximately 160, 120, and 70–90 nucleotides in length, respectively), which make up 15–20% of the total RNA, do not bind in quantity under the conditions used. Thus the size distribution of RNA isolated with the QIA amp procedure is comparable to that of RNA isolated by centrifugation through a CsCl cushion, where small RNAs do not sediment efficiently. Phenol-based procedures co-purify RNAs <200 nt, accounting for the 20% higher yield of RNA purified using phenol-based procedures in comparison to QIA amp-based procedures.

Kit contents:

- Preparations per Kit

- QIAamp Spin Columns (clear)

- QIAshredder Spin Columns (lilac)

- Collection Tubes (1.5 ml)

- Collection Tubes (2 ml)

- Buffer EL*

- Buffer RLT*

- Buffer RW1

- Buffer RPE

- RNase-free Water

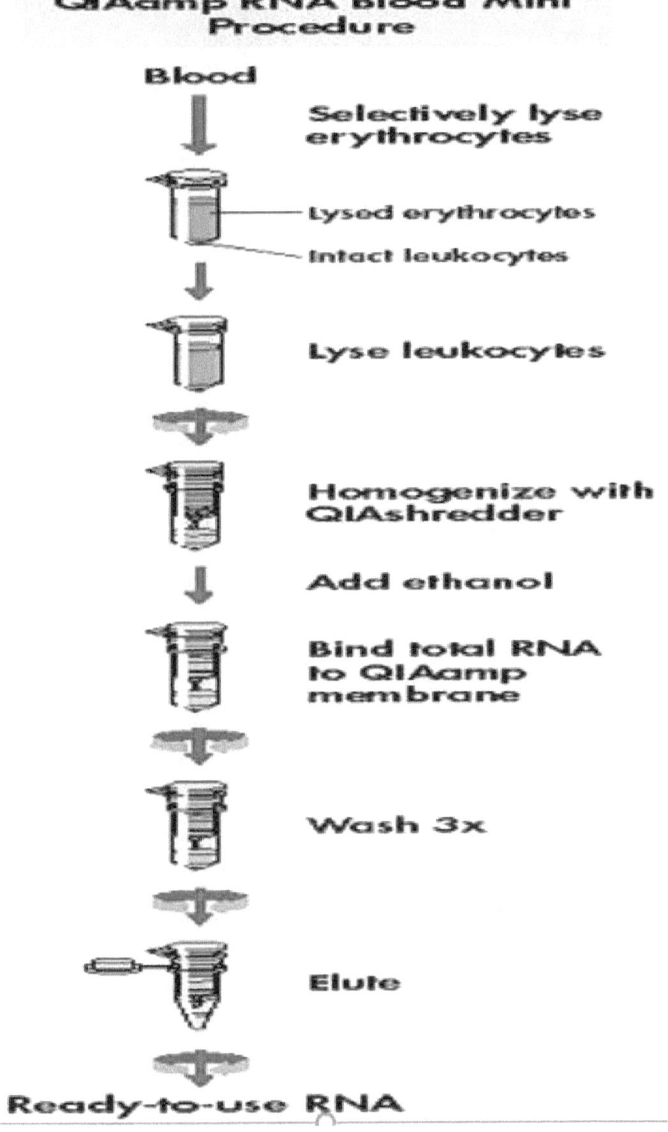
QIAamp RNA Blood Mini Procedure
Blood
Selectively lyse erythrocytes
Lysed erythrocytes
Intact leukocytes
Lyse leukocytes
Homogenize with QIAshredder
Add ethanol
Bind total RNA to QIAamp membrane
Wash 3x
Elute
Ready-to-use RNA

Steps for RNA Isolation from Blood Sample of Bone Marrow

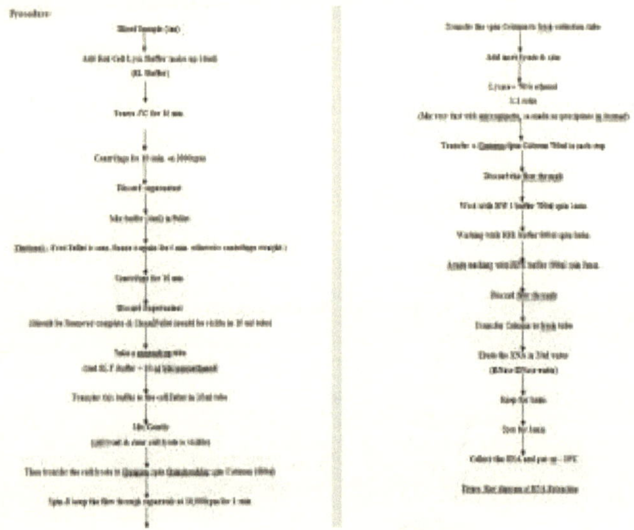

Procedural Steps to conduct RNA Extraction

RT PCR (Reverse Transcription PCR)

RT-PCR is a method used to amplify cDNA copies of RNA (Sambrook & Russell, 2001). For cDNA synthesis 1 ug RNA was dissolved in 8ul of water and mixed with 1 ul random decamers (50uM) and then incubated at 70°C for 5 min and then immediately plunged into ice. This sample mix was then mixed with a cDNA mix containing 4ul Dinucleotide triphosphates (2mM), 1X Reverse Transcriptase Buffer, 40 units Rnase inhibiter and 200 units reverse transcriptase enzyme. The final reaction Mix (20ul) was incubated at 42° C for 60 min and then 93° C for 10 min (for termination of reaction) in a PCR thermal Cycler (M J Research). Negative Control containing sterile water without cDNA was also included.

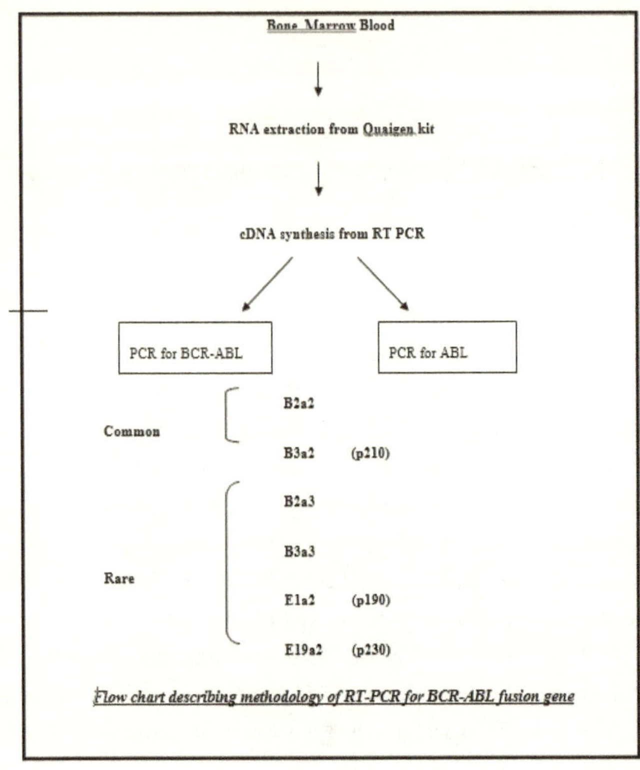

Flow chart describing methodology of RT-PCR for BCR-ABL fusion gene

PCR for BCR-ABL translocation variant

Reagent	Amount	Concentration
Water	9.6 ul	-
MgCl-2(25mM)	2.4ul	4mM

Primer A	1.5ul	300nM
Primer B	1.5ul	300nM
Probe	1ul	200nM
Master Mix	2ul	1X
(Taq polymerase, reaction Buffer, MgCl2, Dntp Mix)		
cDNA	2ul	

PCR (Polymerase Chain Reaction):

PCR is a rapid and versatile in-vitro method for amplifying defined TARGET DNA SEQUENCE present within the source of DNA (Sambrook J.,2001), which need a single copy of a piece of DNA across several orders of magnitude, generating thousands to millions of copies of a particular DNA sequence within a few hours by using a minute amount of DNA. It is a rapid, highly efficient, and sensitive assay for the diagnostic of various diseases like CML.

This technique was developed in 1983 by Kary B Mullis. In 1993, Mullis was awarded the Nobel Prize in Chemistry along with Michael Smith for his work on PCR. Basically, PCR relies on thermal cycling in which we amplify a

TARGET SEQUENCE and undergoes repeated heating and cooling of the reaction for DNA melting and enzymatic replication of the DNA.

PCR REQUIREMENT:

A basic PCR setup requires several components and reagents. (Russel D.,2000)

These components include:

- DNA template that contains the DNA region (target) to be amplified.

- Two oligonucleotide prim hat are complementary to the 3' end of each of the sense and antisense strands of the DNA target.

- DNA polymerase enzyme:-Taq polymerase (extracted from hot spring bacterium Thermus aquatics) with a temperature optimum at around 70 °C.

- Deoxynucleoside triphosphates (dNTPs) are the building blocks from which the DNA polymerase synthesizes a new DNA strand.

- Buffer solution provides a suitable chemical environment for optimum activity and stability of the DNA polymerase

- Divalent cations, magnesium chloride (MgCl2)

RT-PCR size	Amplification conditions Primer	Product sequence
	(Ist amplification	95 C: 1 min
sec	Round) b3a2: 417 bp	94 C: 30
sec	b2a2: 342 bp 5'GAAGTGTTTCAGAAGCTTCTCC 3'	57 C: 60 BCR b1 A
	b3a3: 243 bp 5'GTTTGGGCTTCACACCATTCC 3'	72 C: 60 sec ABL a 3 B
times	b2a3: 168 bp	Go to step II 34
min		72 C: 5
	(IInd amplification	
min	Round) (38)	95 C : 1
		94 C : 30 sec
	b3a2 : 360bp	

57 C : 60 sec   b2a2 : 285 bp   BCR b 2C
5'CAGATGCTGACCAACTCGTGT 3'

72 C : 60 sec   b3a3 : 186 bp   ABL a 3 D
5'TTCCCCATTGTGATTATAGCCTA 3'

Go to step II 34 times   b2a3 : 111 bp

72 C : 5 min

There are three major steps in PCR which are repeated 30 to 40 times. This is done on an automated thermocycler.

The steps are as follows-

- DENATURATION STEP: - It causes melting of the DNA template (92-96oC) by disrupting the hydrogen bonds between complementary bases, yielding single-stranded DNA molecules.

- ANNEALING STEP :-.Synthetically synthesized oligonucleotide primers are added allowing annealing of the primers to the single-stranded DNA template. Stable DNA-DNA hydrogen bonds are only formed when the primer sequence matches the template sequence. The polymerase binds to the primer-template(RNA-DNA) hybrid and begins DNA formation .

- ELONGATION STEP:-The temperature at this step depends on the type of polymerase used.

The following figure demonstrates a flow diagram of PCR.

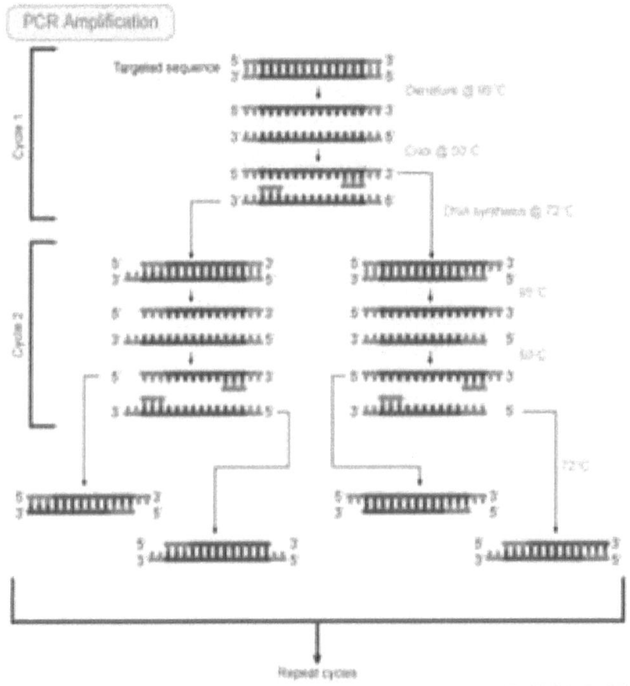

Figure: Different stages of polymerase chain reaction

Gel Electrophoresis Method

Electrophoresis

It is a technique for the separation of the charged molecule through migration, under the influence of the electric field. Agarose Gel Electrophoresis is the technique used to separate larger molecules like DNA in the mixture of DNA fragments on the basis of size and charge. Agarose matrix act as the molecular sieving for the negatively charged DNA molecule (negative charge due to phosphate grp) When the electric current is applied, the larger molecules move more slowly through the gel while the smaller molecules move faster. The different-sized molecules form distinct bands on the gel.

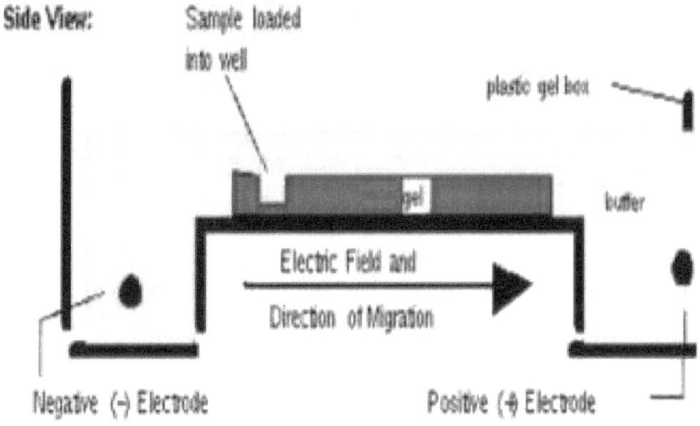

Figure: Schematic representation of Agrose gel electrophoresis apparatus

Visualization of PCR product by AGAROSE GEL ELECTROPHORESIS:-

Following PCR, the amplified product referred to as "amplicon or product" can be detected by performing AGAROSE GEL ELECTROPHORESIS, Where a band containing DNA fragments can indicate the presence of the particular target DNA sequence in the original DNA sample. Products were stained with Ethidium bromide and the gel was visualized under a Gel documentation system (UV transilluminator).

Gel Electrophoresis Requirements

- Agarose (a polysaccharide derivative of agar)

- Electrophoresis Buffer

- TBE buffer (Tris borate EDTA) was used as running buffer, it provides a better resolution resulting in sharp and clean bands. A stock of 20X TBE was used to prepare the final working 1X TBE for the gel run.

Component of 500 ml 20X TBE

Components for 10 X Concentrated	Amount
TRIS base	54 gm
Boric Acid	27.5 gm
0.5 M EDTA	20 ml

Makeup volume up to 500 ml then adjust pH 8 (at pH higher buffering capacity).

To make 500 ml of 1 X TBE:-

- Added 25 ml of 20 X TBE and make the p volume by distilled water up to 500 ml.

- Ethidium Bromide (EtBr) is a mutagen that binds strongly with the DNA by the intercalation between the bases and causes the DNA to fluorescence orange color in the presence of UV light which helps in the visualization of the DNA bands.

- 6 X Loading Dye (Bromophenol Blue) - Loading dye is required for estimating the migration distance of the DNA during the Gel run process.

Components of the 6X loading Dye (Bromophenol Blue)

Components
Concentration

Bromophenol   blue
25mg (0.25%)

Glycerol   3ml

Distilled Water   8ml

DNA ladder /Marker (A set of DNA fragments of known molecular sizes that is used as the standard to estimate the size of an unknown DNA fragment and it runs beside the unknown DNA sample during gel run).

PROCESS OF AGAROSE GEL ELECTROPHORESIS:-

- Preparing 2% GEL: 2 gm of agarose powder was weighed and added to the flask containing 100 ml electrophoresis buffer (1X TBE). The mixture was heated for about 2 mins until the agarose got completely dissolved in the buffer. After that, approximately 7μl of EtBr was added to the mixture

- POURING OF THE GEL: Poured gel mixture in gel casting tray in which combs were already placed. After the solidification, combs were carefully removed and the remaining Depression formed Were "wells".

- The gel casting tray was then kept in gel tank/electrophoresis apparatus and was covered with the running buffer (as the conductor of the electricity and maintains a constant pH).

- The gel should be completely submerged in the running buffer, otherwise, the DNA sample loaded moves out of the well during the gel run.

- LOADING THE GEL: A piece of Parafilm was cut and 3μl of dye was pipette Then 5μl DNA sample was pipette onto the dye 2 μl Ladder of appropriate volume was also pipette into the single well for the quantifying the size of DNA in the simple band, it was differentiated into a distinct band of known sizes Then the gel tank was connected with a power supply by inserting the plug of the black wire into the black input of the power source so that the sample DNA can be able to move from negative to positive size due to negative charge of the DNA.

- After plugging into the power source, the current was applied by applying the power supply ON, Voltage supply of 150V for 15-25 minutes. The flow of current can be confirmed by observing the bubble coming off the electrode i.e. at the cathode side.

- As DNA is negatively charged because of its phosphate group, it will migrate from the negative charge electrode i.e. cathode (Black) to the anode (Red).

- The progress of gel can be monitored by the reference to the ladder

- VISUALIZING THE GEL: Gel was taken out of the casting tray and was placed on the UV transilluminator. UV transilluminator was turned on to see the DNA band and the Ladder.

Finally, the camera of the transilluminator was turned on, and the gel was visualized by the software UV tech, and the gel picture showing the DNA band was taken

Findings and Conclusion

Findings

45

RNA isolation & C-DNA Synthesis:

Successfully isolated RNA from Bone Marrow Blood positive samples using RNA isolation protocol standardized by Qiagen, Germany.

PCR Diagnosis:

BCR ABL TESTS GENE:

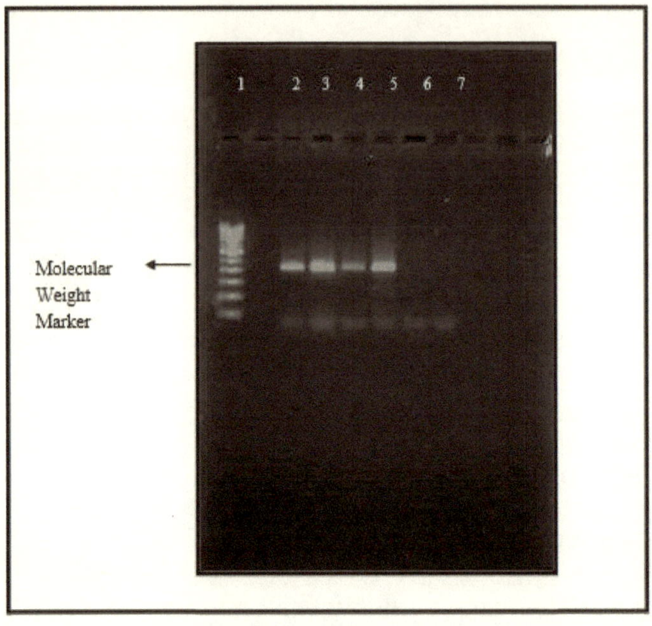

Figure: Ethidium bromide stained agarose gel

RT-PCR products run on Ethidium bromide-stained agarose gels: 1: Molecular weight marker; 2-5: Positive for BCR-ABL PCR 6: Negative for BCR-ABL PCR'7: Negative control

The 5 samples were tested for the presence of BCR ABL Test gene and ABL Control gene using PCR diagnostic assay. Out of 5 samples tested for the identification and confirmation of the BCR ABL Test gene and ABL control gene. 5 were +ve with BCR ABL samples (P1, P2, P3, P4), and the remaining 1 sample (p5) was a negative sample as depicted in the figure.

ABL CONTROL GENE:

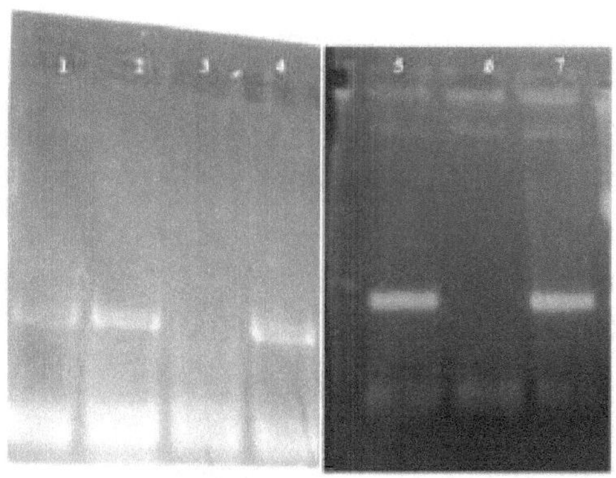

Figure(ii): Ethidium bromide stained agarose gel (ABL Control Gene)

1-5 Patients, sample no. 6 is ABL Control gene -ve and sample no. 7 is ABL Control gene +ve

Conclusion

Discussion

Microscopy is the preferable mean of clinical diagnosis due to its inexpensive and easy availability infield but microscopic identification has been proved to be less sensitive and to detect fusion transcript in CML patients. Bone Marrow cytogenetics is an invasive and time-consuming technique to detect the BCR-ABL fusion gene. PCR diagnostic assay has been proved to be a more robust, specific, and sensitive assay of BCR ABL fusion gene detection. Out of 5 samples 4 tested positive for the BCR-ABL fusion gene. This helped in establishing the diagnosis in these patients while co-relating with other laboratory and clinical findings. Test RT-PCR is a sensitive and less time-consuming technique to detect BCR-ABL transcript to establish the diagnosis in CML patients.

Conclusion

It is true that leukemia is a life-threatening disease and now half of the world's population is at risk of the disease but this disease is curable and preventable. Millions of lives can be saved if the disease is diagnosed early and treated in a more efficient and proper way by the use of effective Imatinib Mesylate drugs.

The PCR used in this study allowed reliable detection of the BCR ABL fusion gene. I have concluded that this PCR is valuable as a confirmatory test and could be implemented to perform the detection of the BCR ABL fusion gene.

Appendix

Abbreviation

ABL	Abelson murine leukemia viral oncogene
ALL	Acute lymphoblastic leukemia
AML	Acute myeloid leukemia
AP	Accelerated phase
ARG	Abelson-related gene
BC	Blast crisis
BCR	Breakpoint cluster region
CBL	Casitas B-lineage lymphoma pro-oncogene
CCgR	Complete cytogenetic response
CHR	Complete hematologic response
CML	Chronic myeloid leukemia
CP	Chronic phase
ERK	Extracellular signal-regulated kinase

FISH   Fluorescence in situ hybridization

GAPDH   Glyceraldehyde-3-phosphate
dehydrogenase

APK   Mitogen-activated protein kinase

MCgR   Major cytogenetic response

mRNA   Messenger ribonucleic acid

NK   Natural killer

Ph   Philadelphia

qRT   PCR Quanlitative reverse transcriptase-
polymerase chain reaction

Bibliography

- Afar D. E., Goga A., McLaughlin J., Witte O. N., and Sawyers C. L., Differential complementation of Bcr-Abl point mutants with c-Myc. Science. 1994;264(5157): 424-6.

- Allan N. C., Richards S. M., and Shepherd P. C., UK Medical Research Council randomised, multicentre trial of interferon-alpha n1 for chronic myeloid leukaemia: improved survival irrespective of cytogenetic response. The UK Medical Research Council's Working Parties for Therapeutic Trials in Adult Leukaemia. Lancet.1995;345(8962): 1392-7.

- Alimena G, Morra E, Lazzarino M, et al. Interferon-μ 2b as therapy for patients with Ph positive chronic myelogenous leukemia : Eur J Haematol.1990;52:25-28.

- Apperley JF .Part I: Mechanism of resistance to imatinib in chronic myeloid leukemia. Lancet oncol.2007; 8:1018-29.

- Arora B, Kumar L , Kumaru M, Sharma A, Wadhwa J & Kochupillai V. Therapy with imatinib mesylate for chronic myeloid leukemia. Ind J Med & Paed Oncology 2005; 26:5-18.

- Aurich J, Duchayne E, Huguet-Rigal F, Bauduer F, Navarro M, Perel Y., Pris J., Caballin M. R., and Dastugue N. Clinical, morphological, cytogenetic and molecular aspects of a series of Ph-negative chronic myeloid leukemias. Hematol Cell Ther.1998; 40(4): 149-58.

- Baccarni M,Sagio G,Goldman J,Hochhaus A,Simonsson B,Appelbaum F, et al. Evolving concepts in the management of chronic myeloid leukemia: recommendations from an expert panel on behalf of the European leukemia Net. Blood.2006;108:1809-182

- Bennett JH (1845). Two cases of hypertrophy of the spleen and liver, in which death took place from suppuration of blood. Edinburgh Med Surg J ;64: 413.

- Ben-Neriah Y, Daley GQ, Mes-Masson A-M, Witte ON, Baltimore D. The chronic myelogenous leukemia-specific P210 protein is the product of the bcr/abl hybrid gene. Science. 1986; 233:212-214.

- Beillard E, Pallisgaard N, Bi W, van der Velden VHJ, Dee R, vander Schoot CE et al. Evaluation of candidate control genes for diagnosis and residual disease detection in leukemic patients using'real-time' quantitative reverse-transcriptase polymerase chain reaction (RQ-PCR) – A Europe Against Cancer Program. Leukemia.2003; 17(12): p. 2474-86.